Laboratory Manual on

Biochemistry

Darani Vasudevan

Preface to the book

Everything needs a beginning. Without basics we could not understand the complicated concepts. The growth in the field of science is mainly due to the methodology and procedures developed by Scientists. Biochemistry has become an invaluable aid in the field of experimental Botany. Although many procedures are being followed, they are found scattered in various books and journals. This book is a collection of various procedures related to Biochemistry. This book would help students, researchers and Scientists of different disciplines in Plant Science. Hope you enjoy reading and find it useful!!!

-V.Darani M.Sc., M.Phil., SET

Preparation of solutions- Percentage, PPM, Molal, Molar and Normality concentrations.

Molar solution

A solution is said to be one molar when it contains one gram molecular weight (1 mole) of a solute in one litre of the solution. Gram molecular weight of solute dissolved in a small quantity of solvent and made upto 1000ml. The final volume should not exceed 1000 ml.

♣0.1 M sucrose solution

Molecular weight of sucrose = 346 g

1M =346 g of sucrose/1000ml Distilled water

   = 34.6 g of sucrose/100ml Distilled water

0.1 M = 3.46 g of sucrose/100ml Distilled water

♣2M NaOH

Molecular weight of NaOH = 40

1M = 40 g NaOH/1000ml Distilled water

   = 4g NaOH/100ml Distilled water

2M = 8g NaOH/100ml Distilled water

Normal solution

A solution is said to be one normal when it contains one gram equivalent weight of a solute in 1 litre of solution.

Equivalent weight= Molecular weight/ Number of replaceable hydrogen

♣0.1 N NaOH

Molecular weight of NaOH= 40g

Equivalent weight = 40/1 = 40g

1N = 40g NaOH/1000ml Distilled water

   = 4g NaOH/100ml Distilled water

0.1 N = 0.4g NaOH/100ml Distilled water

♣2N H_2SO_4

Molecular weight of H_2SO_4 = 82

Equivalent weight = 82/2 = 41

1N = 41g H_2SO_4/1000ml Distilled water

2N = 82g H_2SO_4/1000ml Distilled water

   = 8.2g H_2SO_4/100ml Distilled water

Parts Per Million solution (PPM)

For dilute solutions, it is convenient to express the concentration of the solute in terms of parts per million.

PPM = μg solute/ 10^6 mg water

   = mg solute/ 10^3 g water

   = mg solute/ litre solution

1mg of solute is dissolved in a volume of 1l of solvent.

♣20PPM GA_3

1PPM = 1mg GA_3 / 1000ml distilled water

   = 0.1mg GA_3 / 100ml distilled water

20PPM GA_3 = 20mg GA_3 / 1000ml distilled water

= 2 mg GA_3 / 100ml distilled water

Molal Solution (m)

A solution is said to be one molal when it contains one gram molecular weight (1mole) of a solute and 1000 ml of solvent.

Example: 5m Potassium Dichromate ($K_2Cr_2O_7$)

Molecular weight of $K_2Cr_2O_7$ = 294

1m $K_2Cr_2O_7$ = 294 g+1000ml Distilled water

= 29.4 g + 100 ml distilled water

5m $K_2Cr_2O_7$ = 147g + 1000ml distilled water

= 14.7 g + 100 ml distilled water.

Percentage solution

A solution is said to be 1 % when it contains 1ml of solute and 99 ml of solvent (or) 1g of solute in 100 ml of solvent.

Example: 10% and 1% Sucrose solution

Molecular formula of Sucrose = $C_{12}H_{22}O$

Number of Hydrogen atoms = 22

1% Sucrose = 1g of sucrose/ 100 ml distilled water

0.1% sucrose = 100mg or 0.1g/ 100ml distilled water.

Preparation of Buffers (Phosphate and Citrate Buffers)

Aim

To prepare the standard buffer of phosphate and citrate

Requirements

Monobasic sodium phosphate, dibasic sodium phosphate, citric acid, sodium citrate, HCl, distilled water, measuring jar.

Procedure

Phosphate buffer

Stock solution A

0.2 mol/litre solution of monobasic sodium phosphate (31.2g of $NaH_2PO_4.2H_2O$ in 1 litre)

Stock solution B

0.2 mol/litre solution of dibasic sodium phosphate (28.39 g of Na_2HPO_4 or 71.7g of Na2HPO4.2H2Oin 1 litre). Take X ml of A and Y ml of B, make upto 200ml.

Citrate buffer

Stock solution A

0.1mol/litre solution of citric acid (19.12g in 1litre)

Stock Solution B

0.1mol/litre solution of sodium citrate (29.41g of $C_6H_5O_7Na_3.2H_2O$ in 1 litre). Take X ml of A and Yml of B, made upto 100 ml.

Estimation of Reducing sugar (Nelson- somogyi Method, 1945)

Aim

To estimate the amount of reducing sugar of fresh leaf materials

Requirements

Fresh leaves of *Tridax* and *Millingtonia,* copper reagent, arsenic molybdate colour reagent, distilled water, motor and pestle, test tubes, water bath with heater, pipettes, ice bath, stirrer and spectrophotometer.

Principle

The principles that reducing sugar reduces $CuSO_4$ in the reagent to cuprous oxide (Cu_2O), the cuprous oxide on reduction with arsenomolybdate reagent produces a green coloured complex which is measured at 520nm. The Λmax is 660nm where the sensitivity is 4 times higher than that at 520nm. But 520nm is chosen because sensitivity and the advantages gained by reducing to a minimum, the effect of variation due to reagents reoxidation of cuprous oxide etc. The colour is stable for atleast 18 hours.

$CuSO_4.5H_2O$+Reducing sugar $\longrightarrow$ CuO + sugar (oxidised)

Cu_2O + Arsenomolybdate reagent $\longrightarrow$ Green coloured complex+ CO_2

Procedure

Extraction'1g of fresh leaves of *Tridax* and *Millingtonia* are taken and grind with 10ml of 80% ethanol using pestle nad mortor separately. The extract was centrifuged at 5000 rpm and supernant was taken and made upto known volume 10ml using 80% ethanol.

Assay

Add 2ml of copper reagent to each tube. Place the tubes in a boiling water bath for 10 minutes and then 10 minutes in trough containing tap water. Add 2ml of Arseno molybdate colour reagent to each tube and shaken well. Colour develops very rapidly with the evolution of carbon dioxide.

Read the colour at 520nm using reagent blank. The colour is very stable.

Reagent preparation

I. copper reagent

Solution 1: 1.2g of sodium carbonate anhydrate+ 0.6g of sodium potassium tartarate and made upto 15 ml with distilled water

Solution 2: 200mg of $CuSO_4$/2ml. Distilled water

Solution 3: Solution II+ Solution I and add 0.8g $NaHCO_3$.

Solution 4: 9g of sodium sulphide in 25ml of distilled water; cool and add solution 4 to solution 3 and made upto 50 ml with distilled water.

II. Arsenic molybdate colour reagent

➢ 2.5g of Ammonium molybdate
➢ 45ml of distilled water
➢ Add 2.5ml of conc. Sulphuric acid
➢ Mix well
➢ Add 300mg of Sodium arsenic solution
➢ Dissolved in 1ml of distilled water
➢ Made upto 50 ml.

III. Extraction

➢ 1g fresh leaf material

- ➢ Grind with 5ml of alcohol or water
- ➢ Reextracted with 5 ml of distilled water
- ➢ Centrifuge at 5000rpm
- ➢ Supernant

IV. Estimation

Extract:

- ➢ 2ml of extract
- ➢ 2ml of copper reagent
- ➢ Boil in water bath (10 minutes)
- ➢ Cool in trough containing tap water
- ➢ 2ml of Arsenic molybdate colour reagent
- ➢ Shake well
- ➢ Colour develop rapidly with estimation
- ➢ Absorbancy at 520nm.

Standard (Glucose)

- ➢ 2ml of glucose standard
- ➢ 2ml of copper reagent
- ➢ Boil in water bath (10 minutes)
- ➢ Cool in trough containing tap water
- ➢ 2ml of Arsenic molybdate colour reagent
- ➢ Shake well
- ➢ Colour develop rapidly with estimation
- ➢ Absorbancy at 520nm

Blank

- ➢ 2ml of distilled water
- ➢ 2ml of copper reagent
- ➢ Boil in water bath (10 minutes)
- ➢ Cool in trough containing tap water
- ➢ 2ml of Arsenic molybdate colour reagent

➢ Shake well
➢ Colour develop rapidly with estimation
➢ Absorbancy at 520nm

Standard Graph for starch

S.No	Volume of Stock (ml)	Volume of D.Water (ml)	Total Vol (ml)	Conc. of stock solution (µg)	OD value at 520 nm
1					
2					
3					
4					
5					
6					
7					
8					
9					
10					

OD value of plant Materials

S.No	Name of the plant	parts	OD value
1.			
2.			

Calculation

Fresh weight of the material = 1g

Volume of Extract = 2ml

Conc. of reducing sugar = concentration of starch×OD of plant extract/

OD of starch

Result

The amount of reducing sugar present in the given materials are (μg/fr.wt/g)

Estimation of free Amino acid (Moore and Stein-1948)

Aim

To estimate the amount of total free amino acid present in the given material by Moore and Stein method.

Principle

Ninhydrin, a powerful oxidising agent reacts with all α amino acids between pH 4 and 5 to give the purple coloured compound.

Requirement

Fruit samples of *Ficus benghalensis* and lemon, ethanol, acetic acid, sodium acetate, ninhydrin, standard lysine, distilled water, motor and pestle, test tubes, measuring jar, pipette, centrifuge, beaker and spectrophotometer.

Reagent

➢ 80% ethanol
➢ 0.1M acetic acid and sodium acetate buffer (pH 5.2)
➢ Standard lysine solution
➢ 1% Ninhydrin

Procedure

Extraction

1g fresh materials of *Ficus benghalensis* and lemon are taken and grind with 5ml aqueous solution, 80% ethanol, using a motar and pestle separately. The extract was then centrifuged at 5000rpm and the supernant was made upto known volume (10ml) using distilled water.

Assay

4ml of alcohol and extract was taken in small beakers and the alcohol was allowed to evaporate using sand bath. The residue was made upto 4ml with distilled water and their solution was taken in a separate test tube. To this 0.5ml of sodium acetate buffer was added, followed by 1ml of Ninhydrin solution. The reaction mixture was heated in a water bath at 100°C. Then the samples were allowed to cool at room temperature. Standard solution were prepared taking 4ml of diluted standard lysine solution and all the reaction were carried out as above and absorbency was measured at 660nm in spectrophotometer.

Reagents

➤ 80% ethanol
➤ Sodium acetate buffer (pH 5.2)
➤ 1% ninhydrin
➤ Standard stock solution (10mg/100ml/distilled water)

Extraction

➤ 1g fresh material
➤ Extraction with 5ml 80% ethanol.
➤ Reextract with 5ml 80% ethanol
➤ Centrifuge at 5000 rpm for 15 minutes
➤ Supernant made upto 10 ml with ethanol
➤ Evaporated in sand bath
➤ Residue dissolved in 4ml Distilled water.\

Estimation

Extract

➤ 4ml of extract
➤ 0.5ml Sodium acetate buffer
➤ 1ml Ninhydrin
➤ Mixture heated in water bath at 100°C for 15 minutes.
➤ Allow to cool
➤ OD at 610nm

Standard

➢ 4ml lysine stock solution + 0.5ml sodium acetate buffer + 1ml ninhydrin
➢ Mixture heated in water bath at 100°C for 15 minutes
➢ Allow to cool
➢ OD at 610nm

Standard graph for lysine

S.No	Volume of Stock (ml)	Volume of D.Water (ml)	Total Vol (ml)	Conc. Of standard (µg)	OD value
1					
2					
3					
4					
5					
6					
7					
8					
9					
10					

OD value of plant Materials

S.No	Name of the plant	parts	OD value
1.			
2.			

Calculation

The amount of total free amino acids were expressed as ,

Aminoacid = OD value of sample×Conc. of lysine/ OD value of

standard lysine

Result

The free aminoacid content of one gram fresh materials of-------is (μg/fr.wt/g)

Estimation of Proline (Bates et al., 1973)

Aim

To estimate the free proline of fresh leaf material.

Requirements

Fresh leaves of *Clitoria* and *Tridax*, sulphosalicylic acid, Ninhydringlacial acetic acid, toluene, standard proline, distilled water, motar and pestle, test tubes, water bath with heater, pipettes, ice bat, stirrer, spectrophotometer.

Reagents

➢ Aqueous sulpho salicylic acid (3%).
➢ Acid ninhydrin ⟶ Dissolve 1.25g ninhydrin in a mixture of glacial acetic acid 30ml and 20ml of 6ml phosphoric acid with agitation. The reagent is stable for 24 hours when stored at 4°C.
➢ Glacial acetic acid
➢ Toluene

Procedure

Homogenize 1g of leaf tissue (i.e.) grind in motar and pestle with 10ml of sulpho salicylic acid. Filter the homogenate through whatmann no.2 filter paper. Repeat the extraction and pool the filters.

Pipette 2ml of the filtrate into a test tube. Add 2ml of ninhydrin and 2ml of glacial acetic acid. Incubate for 1 hour at 100°C in a water bath. Transfer the tube to an 0°C bath to terminate the reaction. Add 4ml of toluene and mixed vigorously using a test tube stirrer, for 15-20 seconds. Separate the chromophore containing toluene from the aqueous phase. Cool it to room temperature and measure the absorbency at 520nm. Maintain a reagent blank.

Reagents

- 3% sulphosalicylic acid
- Acid Ninhydrin
- Glacial acetic acid
- Toluene

Extraction

- 1g fresh material
- Homogenize with 10ml 3% sulphosalicylic acid
- Centrifuge at 5000 rpm for 10 minutes
- Supernant made upto 10ml with 3%SSA

Estimation

Extract

- 2ml of extract
- 2ml of acid ninhydrin
- 2ml of glacial acetic acid
- Incubate 1 hour in water bath
- 4ml Toluene is added
- Stir for 20 minutes
- Chromophore solution separated
- OD at 520nm

Standard

- 2ml of sulphosalicylic acid
- 2ml of acid ninhydrin
- 2ml of glacial acetic acid
- Incubate 1 hour in water bath
- Place the test tube in ice bath
- 4ml of toluene is added
- Stir for 20 minutes
- Chromophore solution separated

➤ OD value at 520 nm

Blank

➤ 2ml of distilled water
➤ 2ml of acid ninhydrin
➤ 2ml of glacial acetic acid
➤ Incubate 1 hour in water bath
➤ Place test tube in ice bath
➤ 4ml Toluene is added
➤ Stir for 20 minutes
➤ Chromophore solution separated
➤ OD at 520nm

Standard graph for Arginine

S.No	Volume of Stock (ml)	Volume of D.Water (ml)	Total Vol (ml)	Conc. Of stock(µg)	OD value
1					
2					
3					
4					
5					
6					
7					
8					
9					
10					

OD value of plant Materials

S.No	Name of the plant	parts	OD value
1.			
2.			

Calculation

Determine the proline concentration from a standard prepared with authentic proline basis using the following formula,

(OD value of unknown/ OD value of known)× conc. of standard×V×(1000/ fresh weight)

V = Total volume of the extract = 10ml (after dilution)

Fr.wt = Fresh weight taken for analysis = 1g (1000mg)

Result

The amount of free proline present in the fresh leaf materials of ------ (μg/Fr.wt/g)

Estimation of Protein (Lowry's method)

Aim

To find out the amount of protein in the given plant material by Lowry's method.

Materials required

Fresh leaves of Clerodendron and Millingtonia, Motar and Pestle, distilled water, beaker, measuring jar, pipette and burette, test tubestand, centrifuge and spectrophotometer.

Lowry's method

This is the most widely used method in all laboratories. In this method the colour development relies on the formation of a copper-protein complex as in burette reaction. Cu^{2+} also acts as a catalyst in the reduction reaction.

Reagents

➢ 0.1%NaOH
➢ 4g of NaOH in 100ml distilled water
➢ 10%TCA
➢ 10ml of TCA in 90ml distilled water.
➢ Alkaline copper reagent
➢ Freshly prepared on the day of use by mixing 10ml of 2% $CuSO_4$ and 10ml of 2% potassium sodium tartarate (2g in 100ml each).
➢ BSA

100mg in 100ml distilled water (v/v)

1ml BSA + Distilled water

➢ Folin Phenol (1:1 ratio)

1ml Folin phenol + 1ml distilled water.

Procedure

Preparation of the extract

Measure 1 g of fresh leaf material of Clerodendron and Millingtonia each. Grind them separately with 5ml of distilled water in a pestle and mortar. The extract is taken in a centrifuge tube and centrifuged at 5000 rpm for 10 minutes. After centrifugation process, the supernant is taken and made to 5ml with distilled water. To this add 5ml of 10% TCA.

The solution is incubated at 0°C for 30 minutes. Then it is allowed at room temperature for sometime nad centrifuged at 5000rpm for 5 minutes. The pellet is dissolved in 0.1N NaOH. From this the protein is extracted. The prepared protein extract is stored at 0°C.

Estimation

From the extract, 1ml of protein solution was diluted to 10ml with distilled water. From this 1 ml was taken in a test tube for the blank solution. 1 ml of distilled water was taken in separate test tube. To this add 5ml of alkaline copper reagent. It was allowed to stand in room temperature for 10 minutes. Then 0.5 ml of folin phenol reagent was mixed thoroughly and left standing for 10 minutes at room temperature.

Absorbency was measured at 660nm in systronic spectrophotometer against the reagent. BSA (Bovine Serum Albumin) was used as standard.

Reagents

➢ 0.1N NaOH
➢ 10% TCA
➢ Alkaline copper reagent

- ➢ Folin Phenol
- ➢ BSA

Extraction

- ➢ 1g fresh material
- ➢ Extract with 2.5ml distilled water
- ➢ Reextract with 2.5ml distilled water
- ➢ Centrifuge- 5000rpm for 10 minutes
- ➢ Supernant taken make to 5ml with distilled water + 5ml of 10%TCA (1ml TCA in 9 ml distilled water)
- ➢ Incubation at 0°C for 30 minutes
- ➢ Centrifuge at 5000rpm for 5 minutes
- ➢ Pellet dissolved in 0.1N NaOH
- ➢ Protein extract

Estimation

Extract

- ➢ 1ml of protein solution
- ➢ Diluted to 10ml distilled water
- ➢ Iml sample + 5ml Alkaline copper reagent
- ➢ Allow at room temperature + 0.5ml folin phenol
- ➢ OD at 660nm

BSA standard

- ➢ 1ml BSA standard
- ➢ Diluted to 10ml distilled water
- ➢ 1ml sample + 5ml Alkaline copper reagent
- ➢ Allow at room temperature + 0.5ml Folin phenol
- ➢ OD at 660nm

Blank

- ➢ 1ml water

- ➢ 5ml Alkaline copper reagent
- ➢ Allow at room temperature + 0.5ml of Folin phenol
- ➢ OD at 660nm.

Standard Graph for BSA

S.No	Volume of Stock (ml)	Volume of D.Water (ml)	Total Vol (ml)	Conc. of protein (µg)	OD value at 660 nm
1					
2					
3					
4					
5					
6					
7					
8					
9					
10					

OD value of plant Materials

S.No	Name of the plant	parts	OD value
1.			
2.			

Calculation

Protein content of sample = (OD value of sample × BSA conc.) / OD

value of standard graph

Result

Protein content of ---- (µg/fr.wt/g)

Estimation of Phenol (Mahadevan, 1996)

Aim

To estimate the total phenol content of the plant material.

Principle

Estimation of phenol with folin phenol reagent based on the reaction between phenol and an oxidising agent, phosphomolybdate which results in the formation of blue colour.

Requirements

Leaf materials of Tridax, Clitoria, Pestle and Motar, balance, measuring jar, pipettes, test tubes, stand, distilled water, hot plate and spectrophotometer.

Reagent

- 80% methanol (25ml)
- 20ml methanol + 5ml distilled water
- Folin Phenol (1:1 ratio)

  1ml Folin phenol + 1ml distilled water

- 20% Sodium carbonate

  20g of sodium carbonate / 100ml distilled water

Or

  10g of sodium carbonate /50ml distilled water

- Catechol

1mg in 10ml distilled water

Procedure

Extraction

1g of fresh leaf material was grind with 10 ml of 80% methanol using motar and pestle.

Centrifuged at 6000 rpm for 15 minutes. The supernant was collected and made upto 10ml with distilled water.

Estimation

1ml of extract was taken in a test tube and 1ml of folin phenol reagent (10%) was added to it, followed by 2ml of 20% sodium carbonate solution. The test tubes were shaken well and heated on a boiling water bath for exactly 1 minute. After cooling blue colour solution was obtained and it was diluted to 25ml with distilled water. The absorbancy was measured at 650nm. Phenol standard graph was prepared using catechol.

Reagents

- ➢ 80% methanol
- ➢ Folin phenol reagent
- ➢ 20% sodium carbonate
- ➢ Catechol

Extraction

- ➢ 1g fresh material
- ➢ Ground with 10 ml of 80% methanol
- ➢ Centrifuged at 6000rpm for 15 minutes
- ➢ Supernant made upto 10ml with distilled water

Estimation

Extract

➢ 1ml of extract + Folin phenol reagent (1ml) + 2ml of 20% sodium carbonate
➢ Shake well
➢ Heated in water bath for 1 minute
➢ Allowed to cool
➢ Diluted to 25ml with distilled water
➢ Absorbency at 650nm

Standard

➢ 1ml of catechol + Folin phenol reagent (1 ml) + 2ml of 20% sodium carbonate
➢ Shaken well
➢ Heated in water bath for 1 minute
➢ Allowed to cool
➢ Diluted to 25ml of distilled water
➢ Absorbency at 650nm.

Blank

➢ 1ml of distilled water + Folin phenol reagent (1ml) + 2ml of 20% sodium carbonate
➢ Shaken well
➢ Heated in water bath for 1 minute
➢ Allowed to cool
➢ Diluted to 25ml of distilled water
➢ Absorbency at 650nm.

Standard Graph for catechol

S.No	Volume of Stock (ml)	Volume of D.Water (ml)	Total Vol (ml)	Conc. of phenol (µg)	OD value
1					
2					
3					
4					
5					
6					
7					
8					
9					
10					

OD value of plant Materials

S.No	Name of the plant	parts	OD value
1.			
2.			

Calculation

Phenol content = (OD value of sample × concentration of catechol)/

OD value of standard graph

Result

The concentration of phenol in the given plant material is ----- (µg/fr.wt/g)

Estimation of Ascorbic Acid

Aim

To estimate the ascorbic acid found in the given fruit material.

Requirements

Burette, Pipette, stand, conical flask, electronic balance, oxalic acid, fruit sample, indophenols.

Procedure

500 mg of fruit material was extracted with 90 ml of 4% oxalic acid. The extract was made upto 100 ml with 4% oxalic acid. The supernant was collected and made upto 100ml. The supernant was titrated against the dye.

Dye preparation

➢ 42 mg sodium bicarbonate + 50 ml distilled water + 52mg 2,6 Dichlorophenol indophenol.
➢ Made upto 200ml with distilled water
➢ Stock 10mg Ascorbic acid/100 ml 4% oxalic acid.

Calculation

Amount of ascorbic acid mg/ 100g of sample is,

$(0.5mg/v_1ml) \times (v_2/5ml) \times (100/ \text{weight of the sample}) \times 100$

Result

The ascorbic acid content present in the given fruit material is ---- ($\mu g/g/fr.wt$)

DEMONSTRATIONS

Estimation of oil in seeds

Aim

To estimate the amount of oil present in oil seeds

Principle

Oil from a known quantity of the seed is extracted with petroleum ether, it is then distilled off completely dried off the oil, weighed and the percentage of oil is calculated.

Materials

➢ Peanut
➢ Petroleum ether (40 - 160°C)
➢ Whatman No.2 filter paper
➢ Absorbannt cotton
➢ Soxhlet apparatus

Sample preparation

Take about 50g of kernels in a drying dish and dry at 130°C for not more than 20 minutes in forced draft oven, cool to room temperature and then pass through the nut slicer. Care is to be taken to prevent expressing of any oil while slicing. Mix the sliced sample well. Weigh accurately 2g into the filter paper fold.

Procedure

Fold a piece of filter paper in such a way to hold the seed. Wrap around a second filter paper which is left open at the top. A piece of cotton wool is placed at the top to evenly distribute the solvent as it drops on the sample during extraction. Place the sample packet in the buff tubes in the soxhlet extraction apparatus. Extract with petroleum ether (150 drops/ minute) for pH without interruption (for castor beans use hexane) by gently heating. Allow to cool and dismantle the extraction flask. Evaporate the other or a steam or water bath until no odour of ether remains cool at room temperature.

Carefully remove the moisture outside the flask and weigh the flask. Heating is repeated until constant weight is recorded.

Assay of Amylase or Peroxidase

Aim

To access the enzyme activity per litre of extract

Principle

Guaiacol is used as substrate for the assay of peroxidise

$$\text{Guaiacol} + H_2O \xrightarrow{\boxed{\text{POD}}} \text{oxidized guaiacol} + 2H_2O$$

The resulting oxidized (dehydrogenated) guaiacol is probably more than one compound and depends on the reaction conditions. The rate of formation of guaiacol dehydrogenation product is the measure of the POD activity and can be assayed spectrophotometrically at 436nm

Materials

➢ Phosphate buffer 0.1M (pH 7.0)
➢ Guaiacol solution 20 nm
➢ Dissolved 240 mg guaiacol in water and made upto 100ml. It can be stored frozen for many months. Hydrogen peroxide solution (0.042% = 12.3mm). Dilute 0.4ml of 30% H_2O_2 to 10g MC with water.
➢ The extraction of this solution should be 0.845 at 240nm. Prepare freshly.

Enzyme extract

Extract 1g of fresh plant tissue in 3ml of 0.1M phosphate buffer pH7 by grinding with a precoded mortar and pestle. Centrifuge the chromogenate at 1800rpm at 5°C for 15 minutes. Use the supernant, an enzyme source within 2-4 hours. Store in ice.

Procedure

Pipette out 3ml buffer solution 0.05ml guaiacol solution, 0.1ml enzyme extracts and 0.03ml hydrogen peroxide solution in a cuvet (bring the buffer solution to 25°C before assay). Mix well and place the cuvet in the spectrophotometer. Wait until the absorbency has increased by 0.05. Start a stop watch and note the time required in minutes to measure the increase in the absorbency of the oil.

www.ingramcontent.com/pod-product-compliance
Lightning Source LLC
Chambersburg PA
CBHW071201220526
45468CB00003B/1107